INTRODUCTION TO COMPUTER PROGRAMMING

for kids of all ages

Joe Strout

illustrated by Allie Daigle

MiniScript Press

Printed in the United States of America

First Edition

10 9 8 7 6 5 4 3 2 1

Table of Contents

About this Book

This book is written for two kinds of readers. The first kind is someone who has never before done any computer programming. Sure, you've heard the term, and you probably know that it's how games and apps are made — but you may not know what programming is actually *like*. But by the end of this book, you will!

To further help your understanding, you will find "Programming Puzzles" in yellow boxes throughout the book. These are little challenges to write a bit of code, or change one of the programs in a small way. Answers to these puzzles are given in the back of the book.

You're starting on a very exciting journey. Take your time, enjoy the trip, and above all, have fun!

The other sort of reader is someone who has used other programming languages, but is curious about MiniScript or Mini Micro in particular. Through this book you'll get a very fast and enjoyable introduction to what these tools can do, and be well on your way to creating your own Mini Micro games.

We begin with some instructions on installing Mini Micro, a free app you can use to run all the programs in this book. Then, after a few pages of background information about computers and programming languages, you will be typing your very first computer commands, followed soon by short programs. By the end of the book you will be entering (and understanding!) complete computer games.

I strongly recommend that you have Mini Micro running as you work your way through this book, and **try all the examples for yourself**. Type them in and see what they do. If you make a mistake, that's OK! You will learn far more this way than you would by merely reading about it.

Installing Mini Micro

To run the examples in this book, you will need to install a free program called **Mini Micro**. Mini Micro is a virtual computer that you can program in the same way on any kind of computer (Mac, Windows, or Linux).

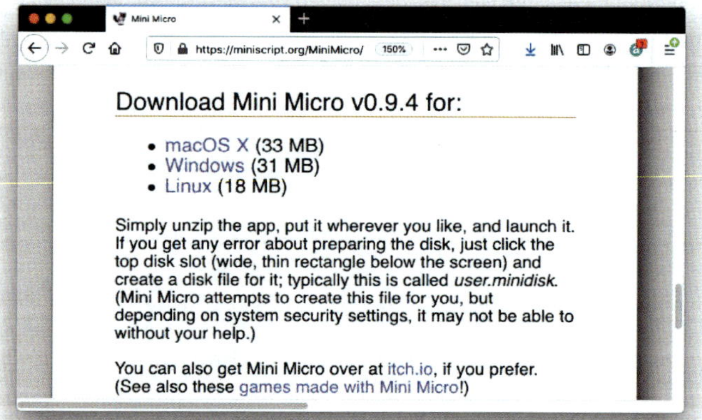

To get Mini Micro, point a web browser to this address:

https://miniscript.org/MiniMicro

Scroll down until you see the **download** links, and click the one that matches the type of computer you have. This should download a file called something like *minimicro-win.zip*. Double-click this to open it up. On Windows, you'll then need to click the **Extract** button; on Mac, it extracts automatically. Then you should find the Mini Micro application.

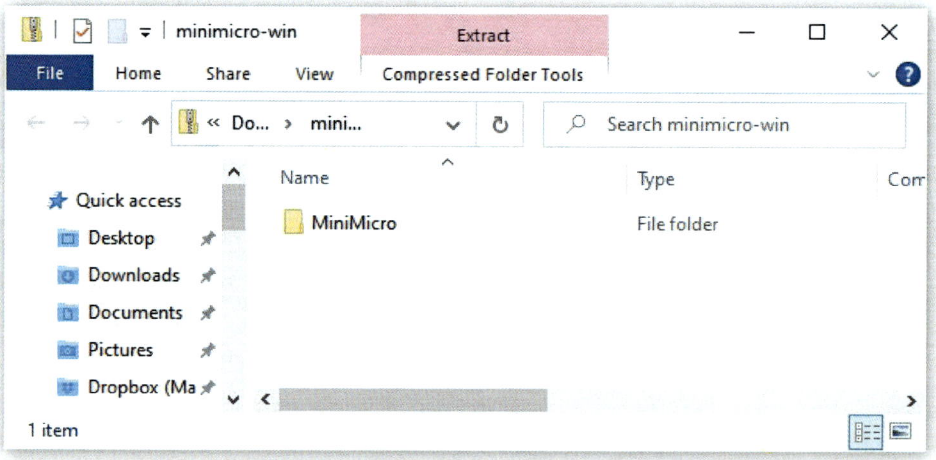

MiniMicro

The Mini Micro application icon looks like the picture at left. Once you find that, double-click it to launch Mini Micro. If you get a security warning, for example about needing access to files, go ahead and allow it. Mini Micro will need to read and write files in order to save your work.

How computers work

Computers are everywhere. There are the obvious ones, like your smart phone or game system. There are less obvious ones, like in your microwave or your bedside clock. And then there are the really invisible ones, like the 50 or so small computers in your car.

All these computers work in pretty much the same way. A central processing unit (CPU) reads an instruction from memory, does what it says, and repeats, billions of times per second.

The instructions at this level are very simple, like comparing two numbers or adding them together. By giving it the right instructions, a computer can be made to do almost anything, from playing Mozart to driving a truck.

It's helpful to think of a computer in terms of its *inputs*, *outputs*, and *processing*.

Inputs are how information gets into the computer. Often this input comes from the user, via a keyboard, mouse, or game controller. On a touch-screen device like a phone or tablet, touch positions are also inputs.

Outputs are things the computer can control, like drawing images on the screen and making sounds with the speakers.

Processing is what the computer actually does, usually to produce outputs that depend in some way on the inputs.

Telling a Computer What to Do

Computers can carry out simple instructions, like adding numbers together, very fast — much faster than humans can. But they don't have any common sense. For example, if someone asks you to give the cat a can of food, you probably know to open it first! But a computer doesn't think of things like that. It has to be told what to do, step by step.

If you had a robot that could understand instructions, and you wanted it to feed and water your cat, you might write out instructions something like this:

1. Go to the cupboard.
2. Get out a can of cat food.
3. Go to the can opener.
4. Open the can with the can opener.
5. Put the can down for the cat.
6. Pick up the water bowl.
7. Go to the faucet.
8. Fill bowl with water from the faucet.
9. Put the bowl down for the cat.

Even this wouldn't be detailed enough for a computer, though. "Open the can with the can opener" would need a whole set of sub-steps explaining how to do that. And while step 5 says to put the can down, it doesn't specify which *way* to put it down. The robot would probably just drop it, making a mess.

The thing about computers is, they're fast, but they're just not very smart. You have to spell everything out for them, and if you have a mistake in your instructions, they will faithfully perform that mistake.

But that's OK. A computer is really just a sophisticated tool. Like any tool, it takes some practice and experience to make it do what you want — but once you do, it can be used to create some really amazing things!

Programming Languages

Natural human languages like English are too imprecise for computers. So, we give instructions to computers using special **programming languages**. These are designed to be unambiguous and precise.

There are many, many different programming languages. They may look different, but they all do the same thing: provide a way to give step-by-step instructions in a format that a computer can understand.

```
evenodd() {
   [[ $(($1%2)) -eq 0 ]] && return "odd"
   return "even"
}
```

UNIX Shell

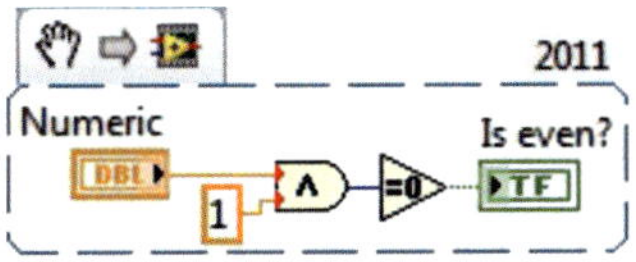

LabVIEW

```
if (x & 1) {
    printf("odd\n");
} else {
    printf("even\n");
}
```

C

```
(defun evenodd (nr)
  (cond ((evenp nr) "even")
   ((oddp nr) "odd")))
```

LISP

```
if x % 2 == 0 then
    print "Even!"
else
    print "Odd!"
end if
```

MiniScript

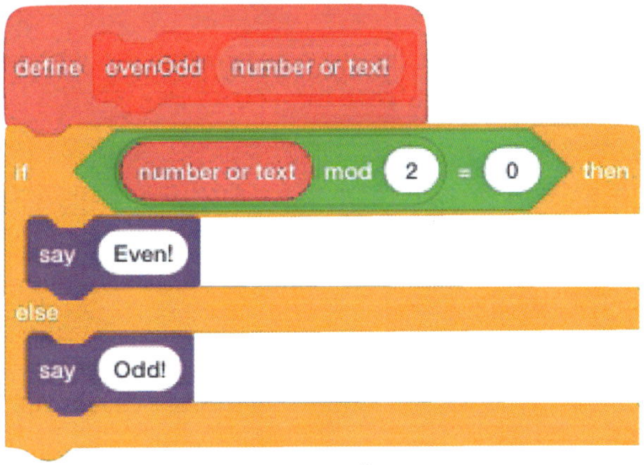

Scratch

A sample of different programming languages is shown above. Some use a lot of funny punctuation, while others use less. Some are made of little pictures you drag around and connect together, but most are made for typing. What other differences can you see?

In this book you'll be using a language called **MiniScript**. It is a fairly new language, and designed to be especially simple and clear. But once you've learned one programming language, it's easy to pick up others if you ever need them.

First Commands

Launch Mini Micro. (If you don't already have Mini Micro installed, please follow the instructions on page 4.) You should see a black screen with a beige border, displaying a welcome message.

After the welcome message is an orange square bracket, and a blinking orange box called the **cursor**. This **prompt** is an invitation for you to type a command. The computer is just waiting for you to tell it what to do!

The real screen is much bigger than this. Or the text is smaller. Anyway, we've tweaked it here so that it is easier to read in this book.

The first command to try prints a message to the screen. Type the following, exactly as shown:

```
print "hello"
```

Be careful to type all in lower case, and to use quotation marks around "hello". Then press the Return or Enter key.

Mini Micro should respond by printing "hello" (without the quotation marks) on the next line. On the line after that, the square bracket and blinking cursor has returned, ready for you to type *another* command.

Computers are fast, but not very smart. And they can be very picky! If you misspell a word, or use a capital letter where it should be lower-case, the computer will not understand what you mean. Instead of doing what you wanted, it will print an **error message**. But that's OK! That's just the computer's way of letting you know it is confused.

Go ahead and try it! Enter "Print" with a capital letter, or misspell it like "pront" and see what happens. Don't worry, you can't hurt anything!

The word `print` is called a **command** or **function**. It's a word the computer knows, if you type it correctly. But the message you want to print can be anything, as long as you put quotation marks around it. Try:

```
print "Hello, world!"
```

Programming Puzzle

Make the computer print the message "Nice to meet you!"

Solutions to programming puzzles begin on page 48.

Your screen may be getting a little cluttered by now. Let's clean it up! Enter this command, exactly as shown, then press the Enter or Return key:

```
clear
```

This command will be very useful any time your screen is full of junk, covered with pictures, or otherwise a mess. Just enter `clear` and you'll be back to a nice clear screen, ready to type your next command.

Variables

Sometimes you need to store something like a number or a message (called a **string** in computer jargon) to use later. In computer programming, you store such values in **variables**. A variable is like a little storage bin for a string, number, or other value.

Each variable has a name, and can store only one thing at a time. When you put a new value into a variable, any previous value stored there is thrown away.

Type the following two lines into Mini Micro, pressing Return or Enter after each one. (Don't type 001 and 002; those *line numbers* are only printed in this book to make it easier to talk about the commands.)

```
001 greeting = "hello"
002 print greeting
```

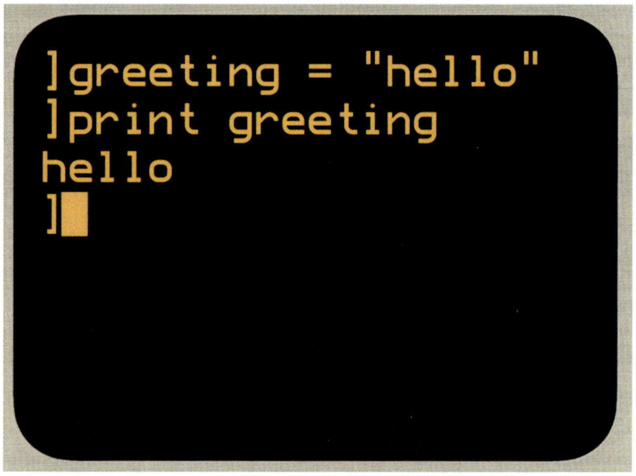

Variables are how computers remember things! Once something is stored in a variable, it stays there until it is replaced by a new value, or cleared by the `reset` command.

Using the Code Editor

So far, you've been typing commands at the Mini Micro prompt. Commands that you enter in this way are carried out immediately. But sometimes you want to list a whole bunch of commands, like a recipe for making a cake. In computers, such a recipe is called a **program**.

Mini Micro has a special kind of text editor (also called a **code editor**) for entering and editing programs. To get to it, type `edit` at the prompt and press Enter or Return.

```
edit
```

The result looks like this, but empty:

Go ahead and type in the code shown above. Then click the green ⊗ button to exit the code editor, and return to the Mini Micro prompt. Now to run your program, type `run` and press Return or Enter.

```
clear
print
print "  |\__/,|   ('\"
print "  |_ _  |.--.) )"
print "  ( T   )     /"
print " (((^_(((/(((_/"
print
```

Try a longer one! Enter `reset` at the prompt to clear the last program, then `edit` to start a new one. Carefully type in the program at left. Press ⊗ to exit the code editor, then `run`. What do you see?

Getting Input from the User

So far, any values we stored in variables were typed right into the program itself. But sometimes we want to ask the user for a value when the program runs. That allows the program to do something different every time, depending on what the user types in.

You can ask the user for a value with the `input` function. It looks like this:

```
x = input("What is x? ")
```

variable name (x) — question ("What is x? ")

The variable name, to the left of the = sign, can be the name of any variable you want to store the user's input in. The question can be any string (message) you want to display to the user. The word `input` must be exactly `input`.

Try the commands shown at right. Change the question or your answer if you like.

```
]x = input("Hmm? ")
Hmm? rainbow pony
]print x
rainbow pony
]
```

Programming Puzzle

Make the computer ask your favorite color, and store the result in a variable called favColor.

The program below uses **input** twice, storing the result in two variables called **name** and **food**. Then it prints them back out to make a suggestion.

To enter this program, first use **reset** to clear the previous program, then **edit** to open the code editor. Type the code shown, then exit the code editor and **run** the program.

```
name = input("Name? ")
food = input("Favorite food? ")
print "Well, maybe you should make a cookbook:"
print
print "Down-home"
print food
print "Recipes"
print "  by"
print name
```

Here's a similar program that makes a silly story. This program also shows you a new trick: you can print several strings on one line, by joining them together with a **+** sign.

Typing these programs in is a great way to learn. Take your time and enjoy each one!

```
name = input("Enter a name: ")
animal = input("A type of animal: ")
adj1 = input("An adjective: ")
adj2 = input("Another adjective: ")
verb2 = input("A verb that can rhyme: ")
noun = input("A noun that rhymes with " + verb2 + ": ")
verb1 = input("A past-tense verb: ")
part = input("A body part: ")
print
print name + " had a " + adj1 + " " + animal
print "Its " + part + " was " + adj2 + " as " + noun
print "And everywhere that " + name + " " + verb1
print "The " + animal + " was sure to " + verb2
```

Numbers and Math

So far, we have mostly worked with strings like "hello" and "recipes." But computers can also work with numbers, like `42` and `-5.73`. In fact, computers are really *good* at working with numbers!

```
]print 38 + 4
42
]print 42 / 7
6
]print 6 * 8
48
]print 48 - 6
42
```

Strings in a computer program must always be typed within quotation marks. Numbers do *not* use quotation marks. That's how the computer can tell the difference between a string and a number.

(If you need to refresh your memory on terms like **string**, see pages 52-53.)

Anything you can do with a calculator, you can do in a computer program. On a computer, these calculations are written using **math operators**: +, −, *, /, and ^.

You can put spaces around the operator, or leave them out. It's up to you.

Operator	Meaning
+	addition
−	subtraction
*	multiplication
/	division
^	power

Programming Puzzle

Use Mini Micro to figure out what you get when you divide 15,750 by 375. **Tip:** don't include the comma when typing big numbers on a computer. Just type `15750`.

You can also add two strings together (as in the silly-story program on page 13), or multiply a string by a number, which just repeats it that many times.

```
print "=" * 68
print " " * 30 + "THE END"
print "=" * 68
```

What does this program do?

Careful! The `input` function always returns a string, not a number. But you can convert a string into a number by using another function called `val`. Used together, you can ask the user for a number like this:

```
x = val(input("Give me a number:"))
```

Try this program which calculates your age in months, days, hours, and minutes!

```
age = val(input("How old are you?"))
print "You're " + age + " years old?"
print "That's " + age*12 + " months"
print "or " + age*365 + " days"
print "or " + age*365*24 + " hours"
print "or " + age*365*24*60 + " minutes!"
```

Programming Puzzle

Add one more line to the program above, that prints your age in seconds.

Making Decisions

Many computer programs need to do different things depending on some value entered by the user. This is done with an **if statement**. Try out this program:

```
001 answer = input("What is 2+2? ")
002 if answer == "4" then
003    print "Wow, humans are smart!"
004 else
005    print "Silly human!"
006 end if
```

Operator	Meaning
==	equal to
!=	not equal to
>	greater than
>=	greater than or equal to
<	less than
<=	less than or equal to

Line 1 asks the user to input an answer. At line 2, the `if` statement checks whether the answer entered is equal to "4". If it is, the computer goes on to line 3, and then jumps down to the `end if` on line 6. But if the answer is *not* equal to "4", then it jumps from line 2 down to the `else` part on line 5.

You can do other kinds of comparisons, too. See the table above for all the different **comparison operators** you can use to compare two values.

Run the program several times, trying out different answers!

```
if condition1 then              ← always required
    ...some code...
else if condition2 then         ← optional...
    ...some other code...
else if condition3 then         ← ...as many as you like
    ...still other code...
else                            ← optional (just one)
    ...last-chance code...
end if                          ← always required
```

An `if` statement can have many parts to deal with different conditions. The first part, after the line with `if` and `then`, must always be there. Then you can have any number of `else if` parts, checking for other conditions. The first of these to match will run its code, skipping the rest. Finally, an `else` part runs if *none* of the conditions were true. Which parts do you see in the program above?

A condition can be a simple comparison, or it can be several comparisons joined with `and` or `or`.

```
fav = input("What's your favorite letter? ")
fav = upper(fav)
print fav * 40
if fav == "A" or fav == "E" then
   print "Why are we shouting?"
else if fav == "R" then
   print "Are you a pirate?"
end if
```

Programming Puzzle

Add two more `else if` parts to the program above: if the user enters "S", tell them to stop hissing; and if they enter "O", ask them if they are a ghost!

Here's a fun program that decides if the user is old enough to see a scary movie. (If you don't like its decision, change the conditions — it's your program!)

Because we want to compare numbers, not messages, we convert the user's input into a number using `val` (as on page 15).

```
print "Now showing: MUTANT ZOMBIES IV"
age = val(input("How old are you? "))
if age < 3 or age > 150 then
   print "Frankly, I don't believe you."
else if age < 13 then
   print "Sorry, you're too young for this one."
else if age < 18 then
   print "Well, I guess you can go with an adult."
else
   print "Enjoy the movie!"
end if
```

Looping

When a computer is programmed to do something multiple times, it is called *looping*. The computer is like a robot running around a racetrack, doing the same thing over and over.

One way to do this is with a *while loop*. A while loop repeats everything between `while` and `end while`, as long as the condition after `while` is true.

Try this while loop, which repeats lines 3-4 as long as `num` is less than 20:

```
001 num = 1
002 while num < 20
003    print num
004    num = num * 2
005 end while
```

```
1
2
4
8
16
```

The condition can be anything. It can even be `true`, which makes an *infinite loop*.

Break out of an infinite loop with the `break` command, or by pressing Control-C.

Programming Puzzle

Change the loop above so that it prints numbers up to 1000.

Here's a fun little guessing game. Change the number in line 1, and have somebody else play; then switch roles. (Come back to this program after you've read page 32, and make the computer pick a number by itself!)

```
001 n = 37          ← revisit this after you read p. 32!
002 guess = 0
003 print "I'm thinking of a number from 1 to 100."
004 while guess != n
005    guess = val(input("Your guess? "))
006    if guess < n then
007       print "Too low!"
008    else if guess > n then
009       print "Too high!"
010    else
011       print "YOU GOT IT!"
012    end if
013 end while
```

Another way to loop is the *for loop*. It repeats the lines of code between `for` and `end for`, but also assigns values to a variable (called the *loop variable*) as it goes. The loop exits when it has gone through all the values in the range.

```
for i in range(1, 5)
```

- `i` — loop variable
- `1` — starting value
- `5` — ending value

```
for i in range(10, 1)
   print i + "..."
   wait
end for
print "BLASTOFF!!!"
```

The `for` loop here counts from 10 down to 1. The `wait` command pauses for 1 second.

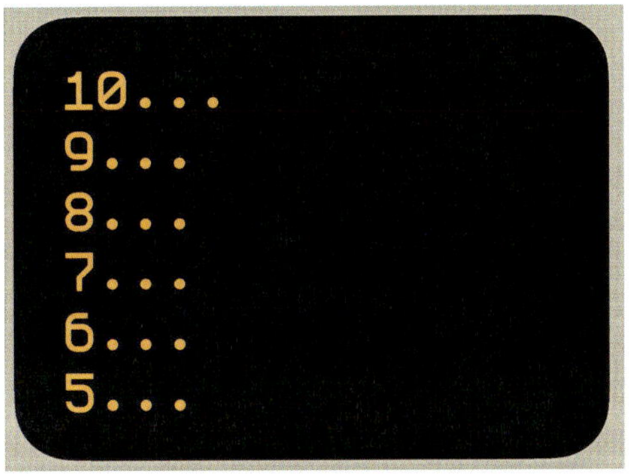

If you need to exit a `for` or `while` loop early, just use the `break` command. `break` jumps immediately to the next line after `end while` or `end for`.

```
for i in range(100, 200)
   print i
   if i == 110 then break
end for
print "Whew, I'm tired!"
```

Programming Puzzle

Change the guessing game on the previous page so it uses a for loop that allows only 5 guesses. Break out of the loop if the user guesses correctly.

```
clear
text.row = 0
print " /\"
print " ||"
print " ||"
print "/__\"
wait
for i in range(1, 25)
   print
   wait 0.3
end for
```

- `clear` — clear the screen
- `text.row = 0` — move cursor to the bottom
- `print` lines 3–6 — print a rocket
- `wait` — wait 1 second
- `for i in range(1, 25)` — repeat 25 times
- `print` — print a blank line
- `wait 0.3` — wait 0.3 seconds
- `end for` — do next repeat

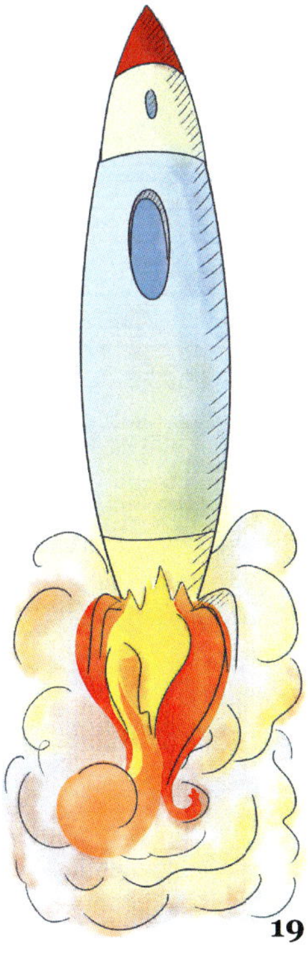

Sounds and Music

Mini Micro can play sounds which are stored in sound files. You can load one using `file.loadSound`, and then play it by using the `play` command.

A number of built-in sounds can be found in the `/sys/sounds` folder. A few of these are shown below; see how many you find when you try it yourself.

```
]dir "/sys/sounds"

   airburst.wav
   aminPluck.wav
   bongo.wav
   bonus.wav
   celloLongC4.wav
   cha-ching.wav
```

The `play` command takes up to three values: the loudness (1 is normal), the left/right pan (0 is centered), and the speed and pitch (normally 1). Try it!

```
snd.play 0.9, 0, 1.5
```

- 0.9: volume (loudness)
- 0: stereo pan: left (-1) to right (1)
- 1.5: speed/pitch

```
]snd = file.loadSound("/sys/sounds/pickup.wav")
]snd.play                  ← just play the sound
]snd.play 0.5              ← play quietly (half volume)
]snd.play 1, -1            ← play it out of the left speaker
]snd.play 1, 1             ← play it out of the right speaker
]snd.play 1, 0, 2          ← play at double speed
```

Changing the speed also changes the pitch, and this lets you make music. A function called `noteFreq` is helpful here; it calculates the sound frequency for any musical note. Note numbers are shown on the piano diagram on the next page.

To calculate the right speed for any note, divide the `noteFreq` of that note by the `noteFreq` of the recorded note (which is usually 60, or middle C).

(That's what's going on in line 10 of the program on the next page.)

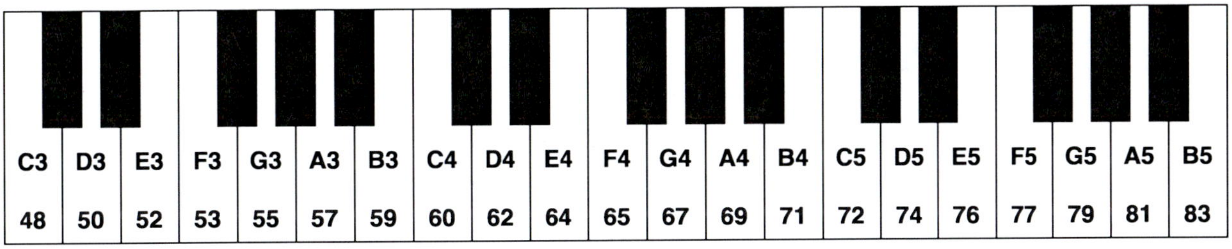

Keyboard Guitar

The program below lets you play guitar on your computer! Use the number keys from 1 to 9 at the top of your keyboard. 1 plays note 60 (middle C), 2 plays note 62, and so on up to 9, which plays note 74.

To find the note number, we first convert the key pressed into a number from 0 to 8 (line 5). Then since most of the white keys on a piano are two notes apart, we multiply this by 2 and add to 60, our base note.

But there is no black key (note) between E and F, or between B and C. So lines 7-8 check for those cases and subtract one for the missing note.

```
001 snd = file.loadSound("/sys/sounds/elecGuitarC4.wav")
002 while true
003    k = key.get
004    if k == char(27) then break
005    num = val(k) - 1
006    note = 60 + num * 2
007    if num > 2 then note = note - 1
008    if num > 6 then note = note - 1
009    print note + " - " + round(noteFreq(note)) + " Hz"
010    snd.play 1, 0, noteFreq(note) / noteFreq(60)
011 end while
```

get key press
break out if Esc pressed
convert "1"-"9" to 0-8
find note number, but
adjust for E-F
and for B-C
repeat

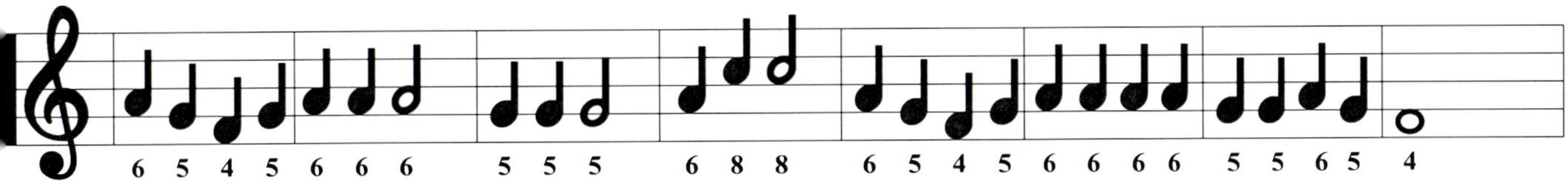

Programming Puzzle

Change this program to play piano instead of guitar! Hint: use `dir "/sys/sounds"` to list all the built-in sounds.

Pixel Graphics

Look at a computer screen with a magnifying glass, and you'll see that it actually displays a bunch of tiny little squares called "pixels" (short for "picture elements"). By changing the colors of these pixels, a computer can display photographs, words, or anything else.

You can draw pixels in Mini Micro by using commands that start with `gfx.`

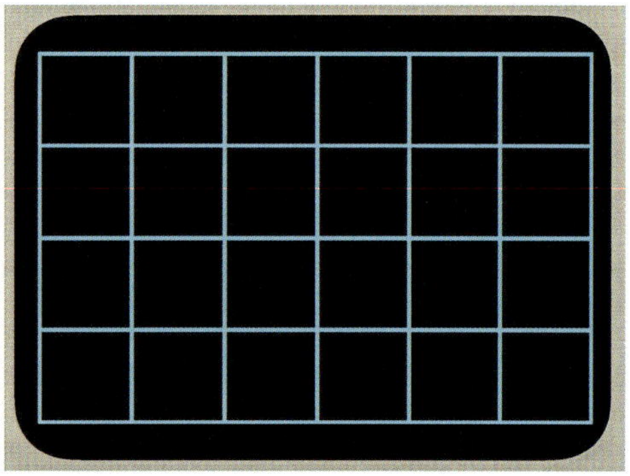

```
gfx.color = color.aqua
for x in range(0, 960, 64)
   gfx.line x,0, x,640
end for
for y in range(0, 640, 64)
   gfx.line 0,y, 960,y
end for
```

set color

draw line

repeat

draw line

repeat

Assign to `gfx.color` to set the drawing color. Use something like `color.aqua`, or a function like `rgb(0, 255, 100)`, where the numbers give the amount of red, green, and blue to go into the color.

Then use the `gfx.line` command with four numbers: x1, y1, x2, and y2. These specify the end points of the line to draw.

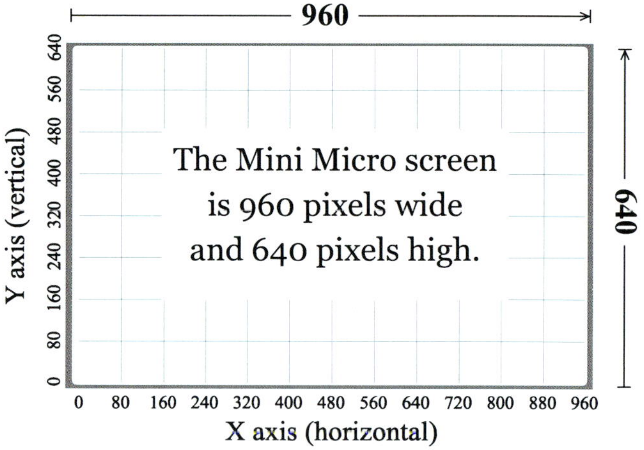

Programming Puzzle

What command would draw a line from the lower-left corner of the screen to the upper-right corner?

```
for x in range(0, 960, 10)
    gfx.color = rgb(x/4, 255-x/4, 200)
    gfx.line 0, 0, x, 640
end for
for y in range(0, 640, 10)
    gfx.color = rgb(200, 255-y/3, y/3)
    gfx.line 0, 0, 960, y
end for
```

You can also use `gfx.fillRect` to fill a rectangle with color. The four numbers it needs are: left side X position, bottom Y position, width, and height. (There's also `gfx.drawRect`, which outlines the rectangle but does not fill it.)

```
for i in range(240, 0, -20)
    gfx.color = rgb(0, i, 0)
    gfx.fillRect 480-i*2, 320-i*1.5, i*4, i*3
end for
```

Programming Puzzle

What do you think will happen if you change `gfx.fillRect` to `gfx.fillEllipse` in the program above? Try it and see!

Cookie Baker

You have 60 seconds to make as many cookies as you can. I made 54,723 cookies. Can you top that?

How to Play	
mouse click	bake 1 cookie
press 1	buy auto-baker for 10 cookies
press 2	buy cookie factory for 100 cookies
press 3	buy cookie company for 1000 cookies

```
clear
cookie = file.loadImage("/sys/pics/food/Cookie.png")
gfx.drawImage cookie, 100, 300, 256,256
cps = 0; cookies = 0; timeLeft = 60
while timeLeft > 0
   text.row = 22
   print "Cookies: " + round(cookies) + "  (" + cps + "/sec)   "
   print "Time: " + round(timeLeft) + "  "
   if mouse.button then
      while mouse.button; end while
      cookies = cookies + 1
   else if key.available then
      k = key.get
      if k == "1" and cookies >= 10 then
            cps = cps + 1
            cookies = cookies - 10
      else if k == "2" and cookies >= 100 then
            cps = cps + 20
            cookies = cookies - 100
      else if k == "3" and cookies >= 1000 then
            cps = cps + 300
            cookies = cookies - 1000
      end if
   end if
   cookies = cookies + cps/60
   timeLeft = timeLeft - 1/60
   yield
end while
```

← main loop
← mouse clicked?
← wait for mouse-up
← bake cookie
← auto-baker!
← cookie factory!
← cookie company!

This program shows a new way to get input from the user, one key at a time. Use `key.available` to see if they have pressed a key, and then use `key.get` to get the letter or number they pressed.

Saving Your Work

After you have typed in a long program, you'd probably like to save it so that it is not lost as soon as you use the `reset` command, or quit Mini Micro. The `save` command does exactly that.

If you've just typed in the Cookie Baker program on the previous page, then at the Mini Micro prompt, enter:

```
save "cookie"
```

This saves the current program to a file called "cookie.ms" on your Mini Micro user disk (which is called `/usr`).

Now, use the `reset` command to clear the current program. Or even quit and relaunch Mini Micro! Then enter:

```
load "cookie"
```

The cookie program you saved before is loaded back into memory. Now you can `run` or `edit` it, just as you did before.

You can name your programs whatever you want. Just remember to always put the program name in quotation marks.

Shortcut #1

You can also save your program while in the code editor, by clicking the  Save button, or by pressing Control-S.

Shortcut #2

If the program in memory has been saved before, you can re-save it in the same place by entering just `save` without a file name.

More File Commands

Mini Micro has two "virtual disks" of files. One is called `/sys`, and contains all the built-in stuff that comes with Mini Micro. The other is `/usr`, and is where you store your own stuff.

You can reference a file on any disk by using a **full path** to the file. That's a string that starts with "/sys" or "/usr", and then gives the name of each folder on the way to the file, and the file itself.

List files with `dir`

The `dir` command lists all the files and folders (directories) under the path you specify. For example:

```
dir "/sys/demo"
```

lists all files in the demo directory on the /sys disk. (There are a lot of fun programs in here — give them a try!)

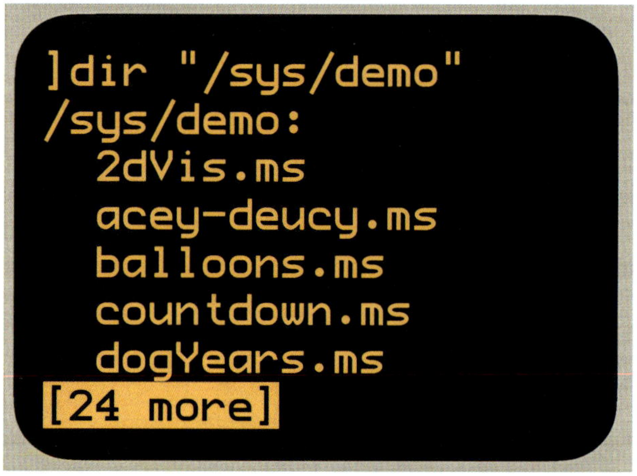

Preview files with `view`

The `view` command displays an image, plays a sound, or lists a program file.

```
view "/sys/pics/Wumpus.png"
```

displays the fuzzy purple monster known as the Wumpus. Or try:

```
view "/sys/sounds/swoosh.wav"
```

Programming Puzzle

How would you load and run the program on the `/sys` disk that draws a sunset? **Hint:** use `dir` to find it!

Most Important Folders	
`/sys/demo`	built-in games and demonstrations
`/sys/pics`	images for use as sprites, etc.
`/sys/sounds`	sound files for you to play
`/usr`	storage for all your own stuff

Mini-Golf Game

Try this fun golf game! You specify how hard you want to hit in terms of speed in X (horizontal) and Y (vertical). Try 5 and 25 for starters. How many swings does it take you to get the ball into the hole?

```
clear
gfx.drawEllipse 900-10, 35-5, 20, 10, color.silver   ← draw hole
gfx.line 905, 35, 905, 65, color.silver              ← draw flag
x = 50; y = 40                                       ← start ball here
gfx.drawEllipse x-5, y-5, 10, 10, color.white        ← draw ball
swings = 1
snd = file.loadSound("/sys/sounds/pop.wav")
while true                                           ← main loop
   print "SWING " + swings
   vx = input("  X speed? ").val                     ← get player inputs
   vy = input("  Y speed? ").val
   snd.play vy/10
   while round(vx) != 0 or round(vy) != 0            ← while ball is moving...
      gfx.drawEllipse x-5, y-5, 10, 10, color.gray   ← erase ball
      vy = vy - 1                                    ← apply gravity
      x = x + vx; y = y + vy                         ← move ball
      if y < 40 then                                 ← if hit ground...
         y = 40
         vy = -vy * 0.7; vx = vx * 0.7               ← bounce!
         snd.play vy/10
      end if
      gfx.drawEllipse x-5, y-5, 10, 10, color.white  ← draw ball
      yield                                          ← wait for next frame
   end while
   if abs(x - 900) < 7 then break                    ← in the hole!
   swings = swings + 1
end while
snd.play 0.5, 0, 0.8
gfx.drawEllipse x-5, y-5, 10, 10, color.gray         ← erase ball
gfx.drawEllipse 900-5, 31, 10, 10, color.white       ← draw ball in hole
print "You got it in " + swings + "."
if swings == 1 then print "A hole in one!  AMAZING!"
```

Lists of Values

Up to this point, a variable has always stored a single value, either a number or a string. But often it is useful to store a whole *list* of values in one variable. To do this, you will need to find the square-bracket keys on your keyboard!

You create a list using square brackets around any number of values, separated by commas.

Type the following to create a list with three elements: "I", "love", and "pi".

```
words = ["I", "love", "pi"]
```

You can also create lists out of numbers, like in this example:

```
primes = [2, 3, 5, 7, 11]
```

Lists may also be created by splitting a string into words, as in the little program below.

In fact there are many handy things you can do with lists — see the table at the bottom of this page.

```
print "Enter a nice long sentence:"
words = input(">").split
words.shuffle
print words.join
```

Useful list-related functions	
lst.push *value*	append *value* to the end of list *lst*
lst.shuffle	randomly reorder the elements of *lst*
lst.sort	sort elements of *lst* into ascending order
lst.len	get the number of elements in *lst*
lst.sum	add up all the elements in *lst*
msg.split	convert string *msg* into a list
lst.join	convert list *lst* into a string
range(1,10)	create a list containing numbers 1 - 10

To refer to any element in a list, you put the *index* of the one you want after the list name, in square brackets. Because computers usually start counting at 0, these indexes start at 0 for the first item, 1 for the second, and so on. You can also count from the end of the list, using negative numbers: -1 is the last item, -2 is the second from the end, etc.

Examples of indexing into a list	
lst`[0]`	first element of list *lst*
lst`[1]`	second element of *lst*
lst`[-1]`	last element of *lst*
lst`[-2]`	next-to-last element of *lst*

```
]arr = [2,3,5,7,11]
]print arr[0]
2
]print arr[-1]
11
]print arr.sum
28
```

Programming Puzzle

Define `arr` as shown at left. What two different ways you could refer to the element with the value of 7 in this list?

Enter the program below to learn some fun facts about the ages in your family.

```
ages = []
print "Enter the ages of your family members"
print "(or 0 when done)."
while true
   age = val(input("Age? "))
   if age == 0 then break
   ages.push age
   print "That's " + ages.len + "..."
end while
ages.sort
print "Your ages are: " + ages
print "Added together, that's " + ages.sum + " years!"
print "The average age is " + (ages.sum/ages.len) + "."
print "The youngest person is " + ages[0]
print "and the oldest is " + ages[-1] + "."
```

Sprites

Mini Micro supports a special kind of drawing called **sprites**. Sprites are little pictures that can move, rotate, and change their size or color, and do it very fast.

You see sprites in computer games like Pac-Man, Super Mario Bros, and Angry Birds. Any time a flat little image moves around on a screen, it is almost always a sprite.

The steps needed to set up a sprite are shown in lines 2-5 of the program below. First you create a sprite with `new Sprite`, and assign it to a variable. Then you load an image for the sprite to show. Next, add it to the display so we can see it with `display(4).sprites.push`. Finally, set its `x` (horizontal) and `y` (vertical) position.

```
001 clear
002 ufo = new Sprite
003 ufo.image = file.loadImage("/sys/pics/UFO.png")
004 display(4).sprites.push ufo
005 ufo.x = 480; ufo.y = 320
006 while true
007    ufo.x = ufo.x + 10 * key.axis("Horizontal")
008    ufo.y = ufo.y + 6 * key.axis("Vertical")
009    yield
010 end while
```

- 001: clear the screen
- 002: create a sprite
- 003: set image
- 004: add sprite to the display
- 005: center it on screen
- 007: update x
- 008: update y
- 009: wait a short time

press Control-C to stop this program!

The while loop on lines 6-10 allows you to move the sprite around. It also shows a new way to get input from the user: `key.axis`. With this method, you can control the sprite with the arrow keys, or the WASD keys, or even a game pad!

Programming Puzzle

Make the UFO move twice as fast! Hint: its current speed is 10 horizontally, and 6 vertically.

Once a sprite is created, you can change its size, color, rotation, or position by assigning to the properties shown in the table at right. See lines 7-8 on the previous page, or lines 11-12 below.

Sprite properties	
spr.`x`	horizontal position of sprite *spr* (0-960)
spr.`y`	vertical position of sprite *spr* (0-640)
spr.`rotation`	rotation angle, in degrees (0-360)
spr.`scale`	size factor: 1=normal, 2=double, etc.
spr.`tint`	sprite color (color.white, color.red, etc.)

Usually you get the image for a sprite by using `file.loadImage` directly. But sometimes the image file on disk is actually a bunch of smaller pictures called **frames**. Each frame represents one step of an animation. To use these, first load the larger image (called a **sprite sheet** in this case) into a variable, then call `getImage` on that to pull out the individual frames, as shown below.

```
001 clear                                              ← clear the screen
002 sheet = file.loadImage("/sys/pics/Enemies.png")    ← load sheet
003 frames = []                                        ← prepare list
004 frames.push sheet.getImage(0, 128*2, 128, 128)     ← grab frame 0
005 frames.push sheet.getImage(0, 128*3, 128, 128)     ← grab frame 1
006 goob = new Sprite                                  ← create a sprite
007 goob.image = frames[0]                             ← set image
008 goob.x = 0; goob.y = 64                            ← position on screen
009 display(4).sprites.push goob                       ← add sprite to the display
010 while true
011    if mouse.x > goob.x + 5 then goob.x = goob.x + 10   ← move right
012    if mouse.x < goob.x - 5 then goob.x = goob.x - 10   ← move left
013    goob.image = frames[(time * 4) % 2]             ← change image to animate
014    yield                                           ← wait a short time
015 end while
```

Frame 0

Frame 1

Rolling the Dice

Many board games are based on rolling dice, which is a way of generating *random numbers*. On each roll, you don't know what you're going to get. Other examples of randomness include flipping a coin or drawing cards from a shuffled deck.

Computers can generate random numbers, too. In MiniScript, you do this with the word `rnd`.

Try the following at the Mini Micro prompt:

```
]rnd
0.642514
]rnd
0.588518
]rnd
0.872987
```

Each time you use `rnd`, it returns a random number between 0 and 1.

```
]rnd * 6
3.640888
]rnd * 6
0.770628
]rnd * 6
5.925875
```

If you multiply `rnd` by 6, then you get a number between 0 and 6.

```
]ceil(rnd * 6)
5
]ceil(rnd * 6)
4
]ceil(rnd * 6)
2
```

Putting `ceil()` around that rounds up, producing a whole number from 1 to 6, like rolling a die.

Programming Puzzles

1. How would you generate a random whole number from 1 to 100?
2. How about a whole number from 51 to 150?

```
001 print "Coin flip!"
002 print
003 for i in range(0,5)
004    text.row = text.row + 1; print "\"; wait 0.1
005    text.row = text.row + 1; print "|"; wait 0.1
006    text.row = text.row + 1; print "/"; wait 0.1
007    text.row = text.row + 1; print "-"; wait 0.1
008 end for
009 if rnd > 0.5 then
010    print "HEADS!"
011 else
012    print "TAILS!"
013 end if
```

The program above flips a coin, using `rnd` on line 9, which will return a value bigger than 0.5 half the time. Everything above that is just for style.

`text.row = text.row + 1` moves the cursor up to the previous line, allowing us to print those slashes, vertical bars, and dashes all in the same place, to look like a spinning coin.

Then try the program below, which makes a worm crawl randomly around on your screen until you press Control-C!

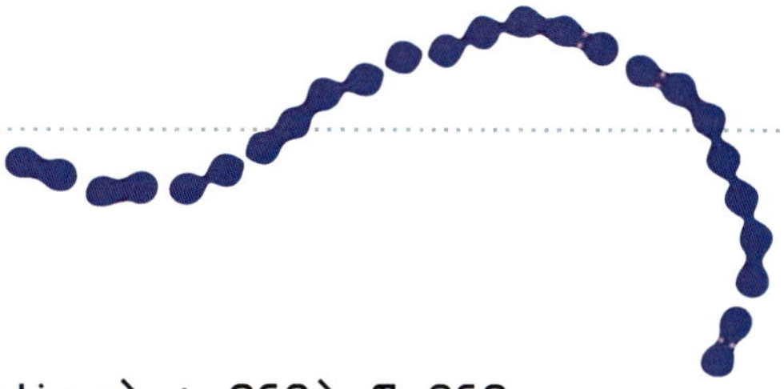

```
xlist = [480]
ylist = [320]
heading = 0
while true
   xlist.push (xlist[-1] + 10 * cos(heading) + 960) % 960
   ylist.push (ylist[-1] + 10 * sin(heading) + 640) % 640
   gfx.fillEllipse xlist[-1]-5, ylist[-1]-5, 10,10, color.aqua
   if xlist.len > 25 then
      gfx.fillEllipse xlist.pull-5, ylist.pull-5, 10,10, color.clear
   end if
   heading = heading + rnd - 0.5
   yield
end while
```

Programming Puzzle

Return to the guessing game on page 18. Change the code so that the computer picks a random number from 1 to 100.

Muffin Catch Game

Here's a game using sprites (page 30) as well as randomness (page 32). Muffins drop from the top of the screen. Use the arrow keys (or a game pad, if you have one) to move the catcher back and forth. Each muffin falls faster than the last. How many can you catch?

This is the longest program in this book so far. Take your time and type carefully (both pages). If you get an error, compare your code to the listing printed here. If you like, save your work with `save "muffin"`, and load it back into memory later with `load "muffin"`.

```
clear
gfx.clear "#DDDDFF"

catcher = new Sprite
catcher.image = file.loadImage("/sys/pics/Block4.png")
catcher.scale = 0.5
catcher.x = 480
catcher.y = 50
display(4).sprites.push catcher

muffin = new Sprite
muffin.image = file.loadImage("/sys/pics/food/Muffin.png")
muffin.x = 480
muffin.y = 500
muffin.speed = 5
display(4).sprites.push muffin

checkedCatch = false
score = 0
muffinCaught = 0
while true
   if muffinCaught then
      muffinCaught = muffinCaught - 1
      if muffinCaught == 0 then   // reset muffin to top
         muffin.y = 720
         muffin.x = round(40 + 920 * rnd)
         muffin.speed = muffin.speed + 1
         checkedCatch = false
      end if
   else  // if muffin is not caught, make it fall
      muffin.y = muffin.y - muffin.speed
   end if
```

← clear screen to a light blue (line `gfx.clear "#DDDDFF"`)

← prepare the catcher sprite... (line `catcher = new Sprite`)

← prepare the muffin sprite... (line `muffin = new Sprite`)

← main loop (line `while true`)

We'll learn about these // things on page 36.

```
   if muffin.y < catcher.y + 42 and not checkedCatch then
      checkedCatch = true
      if muffin.x > catcher.x - 64 and muffin.x < catcher.x + 64 then
            // caught it!
            score = score + 1
            text.row = 25
            text.column = 58
            print "CAUGHT: " + score
            muffinCaught = 60  // hold muffin for 60 frames
      end if
   else if muffin.y < -32 then
      // Muffin fell off the bottom of the screen
      print "GAME OVER!"
      exit
   end if
   dx = 20 * key.axis("Horizontal")
   catcher.x = catcher.x + dx
   if muffinCaught then muffin.x = muffin.x + dx
   yield
end while
```

Programming Puzzle

Add sound effects to this game! **(1)** Play the `pickup.wav` sound when a muffin is caught. **(2)** Play `hit.wav` when you miss. (See page 20.)

Pong Game

Here's another fun game using sprites! This is the classic game of *Pong*, adjusted for one player instead of two. Use the up/down arrow keys (or, again, a game controller if you have one) to move your paddle. Keep the ball on the screen as long as you can!

A line that begins with // (like line 2 in this program) is called a **comment**. A comment is just a little note put in for human readers of the code, like you and me. You can also put a comment at the end of a line, like on line 31. Mini Micro ignores everything from // to the end of the line.

```
001 clear
002 // prepare the ball sprite
003 ball = new Sprite
004 ball.image = file.loadImage("/sys/pics/Block.png")
005 ball.scale = 0.5
006 ball.x = 480; ball.y = 320
007 ball.dx = 10; ball.dy = rnd * 10 - 5
008 display(4).sprites.push ball
009 
010 // prepare the paddle
011 paddle = new Sprite
012 paddle.image = file.loadImage("/sys/pics/Block4.png")
013 paddle.scale = 0.5; paddle.rotation = 90
014 paddle.x = 30; paddle.y = 320
015 display(4).sprites.push paddle
016 
017 // prepare sounds
018 bip = new Sound
019 bip.init 0.2, noteFreq(72), [1,0]
020 bip2 = new Sound
021 bip2.init 0.2, noteFreq(79), [1,0]
022 
023 // main loop
024 lost = false
025 score = 0
026 while ball.x > -30
027    // update the ball
028    ball.x = ball.x + ball.dx
```

← this is a comment! (line 002)

This is just another way to prepare a sound, instead of `file.loadSound`. (lines 018–021)

```
    ball.y = ball.y + ball.dy
    if ball.y < 10 or ball.y > 630 then
       ball.dy = -ball.dy  // bounce off top or bottom
       bip2.play
    end if

    // update the paddle
    paddle.y = paddle.y + key.axis("Vertical") * 10

    // check for ball hitting paddle or back wall
    if ball.x < paddle.x + 30 and not lost then
       if abs(ball.y - paddle.y) < 75 then
             ball.dx = -ball.dx // bounce off the paddle
             ball.dy = ball.dy + rnd * 6 - 3
             bip.play
       else
             lost = true // missed the paddle; game is lost
       end if
    else if ball.x > 930 then
       ball.dx = -ball.dx - 1 // bounce off back wall
       ball.dy = ball.dy + rnd * 6 - 3
       bip2.play
       score = score + 1
       text.row = 25; print score
    end if
    yield
end while
```

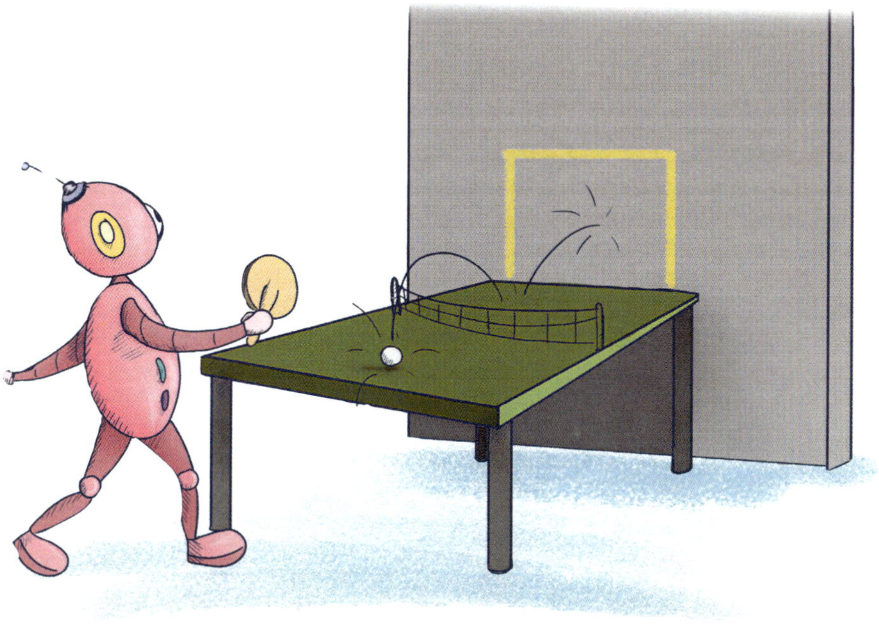

Organizing Code with Functions

The commands you've been using througout this book are properly called **functions**, and the values you specify after the function name are called **parameters** of that function.

For example, `print` is a function that takes one parameter (the string to print). `clear` is a function that doesn't need any parameters. `gfx.line` is a function that needs at least four parameters (x0, y0, x1, and y1), and can take one or two more.

You can define your own functions, too! To do this, use the `function` keyword, followed by variable names for the parameters (if any) within parentheses. In the example at right, the function takes one parameter, and calls it `msg`. Try it! Every time you use the new `say3` command, the code between `function` and `end function` (lines 2-5) runs.

```
001 say3 = function(msg)
002    for i in range(1, 3)
003       print msg
004    end for
005 end function
006
007 say3 "Hello world!"
```

Defining functions is a great tool when you want to do basically the same thing, but with different data. The example below defines a printMeal function that prints out one menu option for a restaurant. Then lines 6-9 call it four times, for four different meals.

```
001 printMeal = function(prefix, main, spamCount)
002    parts = ["spam"] * spamCount + [main]
003    print prefix + ". " + parts.join(", ") + ", and spam"
004 end function
005 print "OUR MENU:"
006 printMeal 1, "eggs", 2
007 printMeal 2, "baked beans", 3
008 printMeal 3, "grits", 5
009 printMeal 4, "steak", 1
```

Remember, the code inside the function (like lines 2-3 in the example above) does not run when the function is being defined. It only runs when the function is *called*, by using the name it was assigned to on the function line.

Here's a program that goes at warp speed! Each of the "warp" lines is represented by a list with five elements: the x0, y0, x1, and y1 coordinates of the line, plus the color. We need to reset these lines when the program begins, and again whenever one goes off the top or bottom of the screen. So, we define a resetLine function to make this easy!

```
001 resetLine = function(m)                          ← define a function...
002    m[0] = 100 * rnd - 50                         ← pick random X0
003    m[1] = 100 * rnd - 50                         ← pick random Y0
004    m[2] = m[0]/2                                 ← start X1 at half of X0
005    m[3] = m[1]/2                                 ← start Y1 at half of Y0
006 end function
007
008 clear
009 lines = []                                       ← prepare a list of lines
010 for i in range(50)                               ← loop 51 times (from 50 to 0)
011    m = [0,0,0,0]                  ← initial coordinates (and random color)
012    m.push color.rgb(155+100*rnd, 155+100*rnd, 155+100*rnd)
013    resetLine m                                   ← call our resetLine function
014    lines.push m                                  ← push onto our lines list
015 end for
016 while true                                       ← repeat until Control-C
017    clear                                         ← clear the screen
018    for m in lines                                ← loop over all our lines
019       gfx.line 480+m[0],320+m[1], 480+m[2],320+m[3], m[4]   ← draw!
020       for i in range(0,3)                        ← for each coordinate...
021             m[i] = m[i]*1.1                      ← make it 10% bigger
022       end for
023       if m[3] > 320 or m[3] < -320 then          ← if off screen...
024             resetLine m                          ← call our resetLine function
025       end if
026    end for
027    yield                                         ← wait for next frame
028 end while
```

Programming Puzzle

Make the warp lines twice as thick! **Hint:** enter `@gfx.line` to see all the parameters that function can take, then make line 19 a bit longer.

Graphing Data

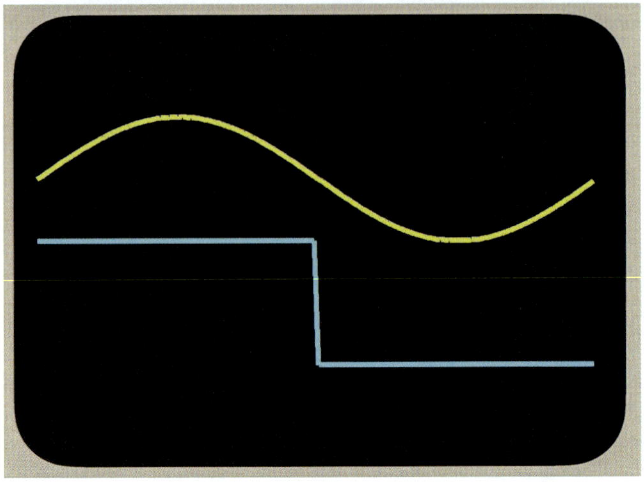

Computers are great at making charts and graphs. Usually you'll use a pixel display like `gfx` for this (see page 22).

If your data is a long list of values (page 28), then make a `for` loop to iterate over that data, calculating the `x` and `y` coordinates on screen from the loop variable and the data value. Use a couple of variables (`lastx` and `lasty` in this example) to keep track of the position of the last point. Draw a line from the last point to the new one, and then continue until the loop is done.

```
plot = function(data, left, middle)
   for i in range(0, data.len-1)
      x = left + i * 7
      y = middle + data[i] * 100
      if i > 0 then gfx.line lastx, lasty, x, y
      lastx = x
      lasty = y
   end for
end function

clear
gfx.color = color.yellow
plot Sound.sineWave, 30, 400
gfx.color = color.aqua
plot Sound.squareWave, 30, 200
```

define a function...
loop over indexes into data
calculate plot position
draw line
remember plot position
plot sineWave in yellow
plot squareWave in aqua

```
clear
counts = [0,0,0,0,0,0,0,0,0,0,0,0,0]  // 12 zeroes

drawBarGraph = function()
   gfx.clear
   for i in range(2, 12)                  ← loop over all possible sums
      x = 70 * i                          ← calculate bar position on screen
      count = counts[i]
      gfx.color = rgb(count*9, count*9, 255)
      gfx.fillRect x, 60, 56, count * 20          ← draw bar
      text.row = 1; text.column = 5 * i + 1  ┐
      print i                                ┘── print label
   end for
end function

while true
   d1 = ceil(rnd * 6)   ┐
   d2 = ceil(rnd * 6)   ├── roll two dice and find the sum
   sum = d1 + d2        ┘
   text.row = 25; print " " * 68             ← clear top row
   text.row = 25; text.column = 5 * sum - 1  ┐
   print d1 + "+" + d2 + "=" + sum           ┘── print result
   counts[sum] = counts[sum] + 1       ← update count for this sum
   drawBarGraph                        ← call function above
   if counts[sum] > 26 then break      ← stop when any count > 26
   wait 0.25                           ← pause briefly to let the user see
end while
```

Another kind of chart is the *bar chart*, where you draw a rectangle (bar) sized to represent each data point. You can tweak the colors too if you want to get fancy.

When the bars represent counts of some event, like the result of rolling two dice, then the chart is called a *histogram*.

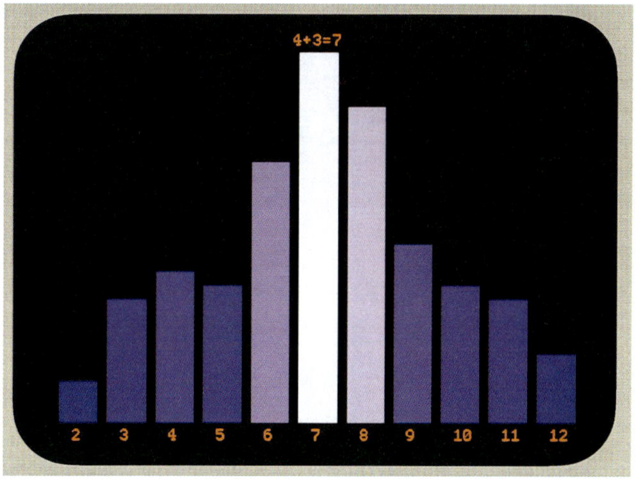

Programming Puzzle

Each of the two programs in this section uses a function to organize the code. What are their names?

Maps and Objects

When you assign values to variables in MiniScript, the computer stores them in a little table in memory. This table maps each variable name to its value. Use the `reset` command to clear memory, then enter the following three assignments.

```
]a = "Hi"
]b = 3.14
]c = 6*7
```

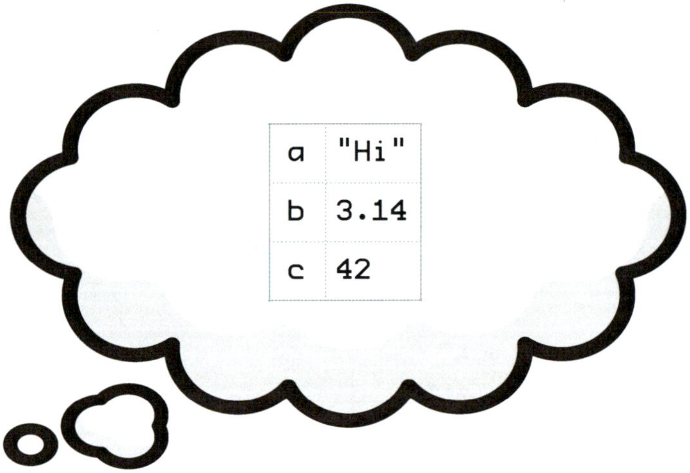

The computer's memory contains a table mapping `a` to `"Hi"`, `b` to `3.14`, and `c` to `42` as shown above. Now, replace the value of b with the new stuff shown below.

```
]b = {}
]b.uno = 1
]b.dos = 2
]b.hola = "Hi!"
```

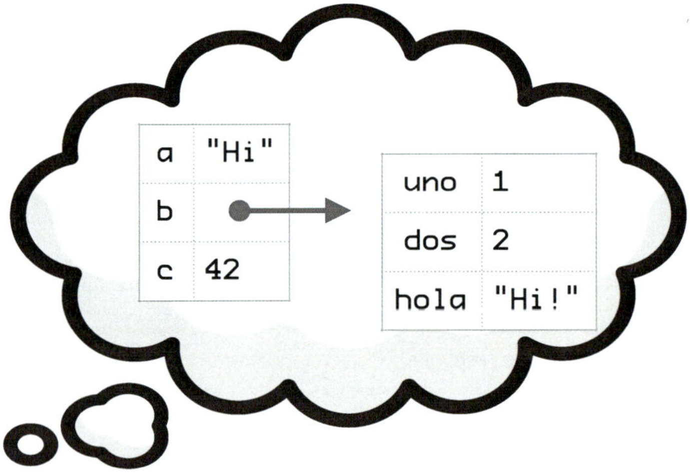

The two curly braces, `{}`, define an empty **map** — a new table of names and values. Our variable `b` now just points to this new table. Using **dot syntax**, like `b.uno`, we can get to the variables inside that map.

After those assignments, the computer's memory looks like the above. Variables `a` and `c` hold simple values, but `b` holds a pointer to another map containing `uno`, `dos`, and `hola`.

You can use dot syntax to get the values back out, too. Try `print b.hola` to print a greeting, right out of the map you just created. It works exactly like normal variables, but with (in this case) a `b.` prefix.

Programming Puzzle

Make `pool` refer to a map that maps `width` to `8` and `length` to `12`. Then use this to print out the area (width times length) of the pool.

You've seen this sort of dot syntax before. Remember this code from page 20?

```
snd = file.loadSound("/sys/sounds/pickup.wav")
snd.play
```

`file` is actually a map that contains file-related functions like `loadSound`. And, what it returns is another map representing the loaded sound, with functions on it like `play`. Sometimes a map like that is called an **object**.

An object can also be created by using `new`, as in `new Sprite`. The `new` command creates a special sort of map that is linked to the map it was created from. In the program below, we create a `Friend` map from `Sprite`, then create `mochi` and `wumpus` from `Friend`.

```
clear
Friend = new Sprite
Friend.scale = 0.5
Friend.y = 320

mochi = new Friend
mochi.image = file.loadImage("/sys/pics/Mochi.png")
mochi.x = 200
display(4).sprites.push mochi

wumpus = new Friend
wumpus.image = file.loadImage("/sys/pics/Wumpus.png")
wumpus.x = 700
display(4).sprites.push wumpus
```

Any values not defined on `mochi` and `wumpus` (like `scale` and `y`) are looked up in `Friend` instead. And because `Friend` was created from `Sprite`, they all get `Sprite`'s functionality too.

Light Bikes Game

```
clear
gfx.clear color.clear
gfx.drawRect 0,0, 960,640, color.aqua, 20

Bike = new Sprite
Bike.image = file.loadImage("/sys/pics/shapes/CircleThin.png")
Bike.scale = [0.5, 0.25]
Bike.y = 320
Bike.speed = 3
Bike.xAhead = function(dist)
   return self.x + cos(self.rotation * pi/180) * dist
end function
Bike.yAhead = function(dist)
   return self.y + sin(self.rotation * pi/180) * dist
end function
Bike.update = function()
   newx = self.xAhead(self.speed)
   newy = self.yAhead(self.speed)
   gfx.line self.x, self.y, newx, newy, self.tint, 6
   self.x = newx
   self.y = newy
   self.speed = self.speed + 0.01
   if gfx.pixel(self.xAhead(16), self.yAhead(16)) != color.clear then
      self.explode
   end if
end function
Bike.turn = function(newDir)
   if abs(self.rotation - newDir) == 180 then return
   self.rotation = newDir
   self.speed = 3
end function
Bike.explode = function()
   gfx.color = self.tint
   for ang in range(35)
      self.rotation = self.rotation + 10
      d = 40 + 150 * rnd
      gfx.line self.x, self.y, self.xAhead(d), self.yAhead(d)
   end for
   file.loadSound("/sys/sounds/airburst.wav").play
   exit  // ends the game
end function
```

Here's an exciting game for two players! One player uses the W, A, S, and D keys to drive the orange bike; the other player uses the arrow keys to drive the green one. If you hit a wall or a bike trail, it's game over! The last bike running wins the game.

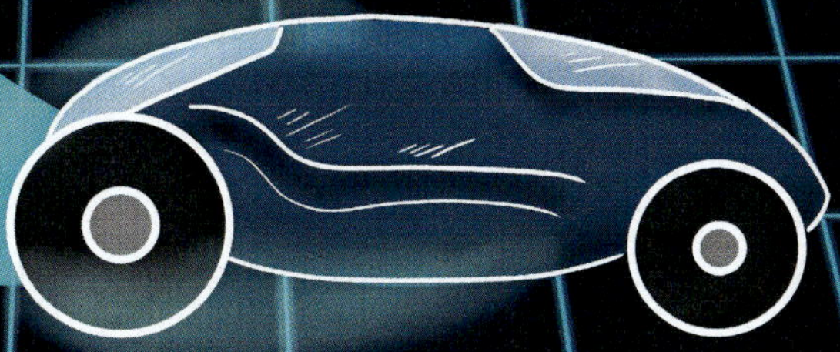

```

players = [new Bike, new Bike]
players[0].tint = color.orange
players[0].x = 480-200
players[0].rotation = -90
players[1].tint = color.lime
players[1].x = 480+200
players[1].rotation = 90
for p in players; display(4).sprites.push p; end for

while true
   for p in players; p.update; end for
   if key.available then
      k = key.get
      // Player 0 uses WASD.
      if k == "a" then players[0].turn 180
      if k == "d" then players[0].turn 0
      if k == "w" then players[0].turn 90
      if k == "s" then players[0].turn -90
      // Player 1 uses arrow keys: char(17) - char(20)
      // for left, right, up, and down arrow.
      if k == char(17) then players[1].turn 180
      if k == char(18) then players[1].turn 0
      if k == char(19) then players[1].turn 90
      if k == char(20) then players[1].turn -90
   end if
   yield
end while
```

Programming Tips

Everybody makes mistakes when writing code. Here are some of the most common ones, and how to fix them! When your code doesn't work, check these to see if any of them might be the problem.

Incorrect capitalization or spelling. For example, if you type `Print` or `pront` instead of `print`, the computer will not understand what you mean.

Using = where you need ==. Remember that to see if two values are equal, you have to use `==`. A single `=` is only used to assign a new value to a variable. You'll also get this error if you try to use `=<` instead of `<=` for "less than or equal to", or `=>` instead of `>=` for "greater than or equal to."

Mismatched block starter and ender. An `if` block must always end with `end if`. Similarly, a block that starts with `while` must end with `end while`, and every `function` needs an `end function`. If you mix these up, the computer will get confused.

Missing quotation marks. Every string in code needs to have quotation marks around it. Be sure to use double quotation marks like "this", not the single kind like 'this'.

If you need to put quotation marks *in* a string, type them twice:

```
print "Say ""hi"" to me."
```

Missing commas. Any command that takes multiple parameters, such as `gfx.line`, must have commas in between them.

Treating a string like a number. Remember that when you get input from the user with `input`, it is a string, not a number. To use it as a number, you need to convert it with `val()`. (See page 15.)

General Advice

Read sample code. But don't *just* read it. Type it in, tinker with it, play with it. Coding is a hands-on activity! It's very hard to truly understand something from only reading it.

Learn by doing. Try taking some of the examples in this book a bit farther. Or create a new program from scratch. If one idea turns out to be too hard to do, pick something else. You will learn and grow from every project.

Check your assumptions with `print`. If your code isn't doing what you think it should, add `print` commands to check the values of your variables.

When stuck, ask for help. Try to figure it out yourself for 10 or 15 minutes, but if still stuck, ask! (And someday you can help others in return.)

Take breaks. Both when debugging, and when just writing code or absorbing new material, a five minute break every half hour recharges your batteries.

Use multiple sources. Learn from other books, the wiki, videos, and so on.

Believe in yourself. Don't get discouraged when coding seems hard. Every programmer struggles sometimes, especially at first. Hang in there, and know that it will get easier!

Programming Puzzle Solutions

Page 9: Type the following:

```
print "Nice to meet you!"
```

and press the Return or Enter key.

Page 12:

```
favColor = input("What's your favorite
color?")
```

(Be sure to type that all on one line.)

Page 14:

```
print 15750 / 375
```

Page 15: `edit` the program, and add:

```
print "or " + age*365*24*60*60 +
" seconds!"
```

(Be sure to type that all on one line.)

Page 17: insert before line 8:

```
else if fav == "S" then
     print "Stop hissing!"
else if fav == "O" then
     print "Are you a ghost?"
```

Page 18: change line 2 to read:

```
while num < 1000
```

Page 19: several changes are needed for this one. First, change line 4 to read:

```
for i in range(1, 5)
```

(Or instead of `i` you could use any variable name not already in use.) Then change the last line, line 13, to:

```
end for
```

Finally, insert one more line before the `end if` on line 12:

```
        break
```

Page 21: in line 1, change `elecGuitarC4` to `pianoLongC4`. Keep the rest of the line the same.

Page 22:

```
gfx.line 0,0, 960,640
```

Page 23: you should see a round tunnel instead of a rectangular one.

Page 26:

```
load "/sys/demo/sunset"
run
```

Page 29:

```
print arr[3]
print arr[-2]
```

Page 30: on line 7, change the `10` to `20`; and on line 8, change the `6` to `12`.

Page 32:

1. `ceil(rnd * 100)`
2. `ceil(rnd * 100) + 50`

Page 33: change line 1 of this game to:

```
n = ceil(rnd * 100)
```

Page 35: insert two new lines before the `while` loop on line 21:

```
caughtSnd = file.loadSound("/sys/sounds/
pickup.wav")
missSnd = file.loadSound("/sys/sounds/
hit.wav")
```

(Be sure not to break the lines; they only appear broken above because of printing limitations.) Then, right after what's now line 42 that prints "CAUGHT," insert:

```
    caughtSnd.play
```

Finally, after the line (now 48) that prints "GAME OVER", insert:

```
    missSnd.play
```

Page 39: to the end of line 19, add a comma followed by a number `2`. (The sixth parameter to `gfx.line` specifies the line width.)

Page 41: the function in the program on page 40 is called `plot`. In the program on page 41, it's `drawBarGraph`. But these are just variable names; you could call them something else, as long as you use the same name where the function is called!

Page 43:

```
pool = {}
pool.width = 8
pool.length = 12
print pool.width * pool.length
```

Character Chart

7	bell	48	0	73	I	98	b	123	{	207	Ï	232	è	964	τ
8	BS	49	1	74	J	99	c	124	|	208	Đ	233	é	8592	←
9	tab	50	2	75	K	100	d	125	}	209	Ñ	234	ê	8593	↑
13	LF	51	3	76	L	101	e	126	~	210	Ò	235	ë	8594	→
17	←	52	4	77	M	102	f	134	inv	211	Ó	236	ì	8595	↓
18	→	53	5	78	N	103	g	135	~~inv~~	212	Ô	237	í	9744	☐
19	↑	54	6	79	O	104	h	161	¡	213	Õ	238	î	9745	☑
20	↓	55	7	80	P	105	i	172	¬	214	Ö	239	ï	9746	☒
21	⍰	56	8	81	Q	106	j	181	µ	215	×	240	ð	9824	♠
32		57	9	82	R	107	k	191	¿	216	Ø	241	ñ	9827	♣
33	!	58	:	83	S	108	l	192	À	217	Ù	242	ò	9829	♥
34	"	59	;	84	T	109	m	193	Á	218	Ú	243	ó	9830	♦
35	#	60	<	85	U	110	n	194	Â	219	Û	244	ô	9856	⚀
36	$	61	=	86	V	111	o	195	Ã	220	Ü	245	õ	9857	⚁
37	%	62	>	87	W	112	p	196	Ä	221	Ý	246	ö	9858	⚂
38	&	63	?	88	X	113	q	197	Å	222	Þ	247	÷	9859	⚃
39	'	64	@	89	Y	114	r	198	Æ	223	ß	248	ø	9860	⚄
40	(	65	A	90	Z	115	s	199	Ç	224	à	249	ù	9861	⚅
41	)	66	B	91	[	116	t	200	È	225	á	250	ú	57856	
42	*	67	C	92	\	117	u	201	É	226	â	251	û	57857	
43	+	68	D	93	]	118	v	202	Ê	227	ã	252	ü	57872	
44	,	69	E	94	^	119	w	203	Ë	228	ä	253	ý	57873	
45	-	70	F	95	_	120	x	204	Ì	229	å	254	þ	57874	
46	.	71	G	96	`	121	y	205	Í	230	æ	255	ÿ	57875	
47	/	72	H	97	a	122	z	206	Î	231	ç	960	π	57888	

Special characters

bell: plays a bell sound.
BS: backspace (moves cursor back).
tab: advance to next tab stop.
LF: linefeed (go to next line).
inv: turn on inverse mode.
~~inv~~: turn off inverse mode

```
text.delimiter = ""
while not key.available
   print char(57873) + char(8); wait 0.3
   print char(57874) + char(8); wait 0.3
end while
text.delimiter = char(13); print
```

Circles, Angles, and Coordinates

There are two different ways to measure rotations: degrees and radians. Degrees go from 0 to 360. Rotating halfway around is 180 degrees, often written with a little circle, like 180°.

Radians go from 0 to 2*pi (pi is just a special number equal to about 3.14). Rotating halfway around is pi radians. Degrees and radians are just two ways of expressing the same idea.

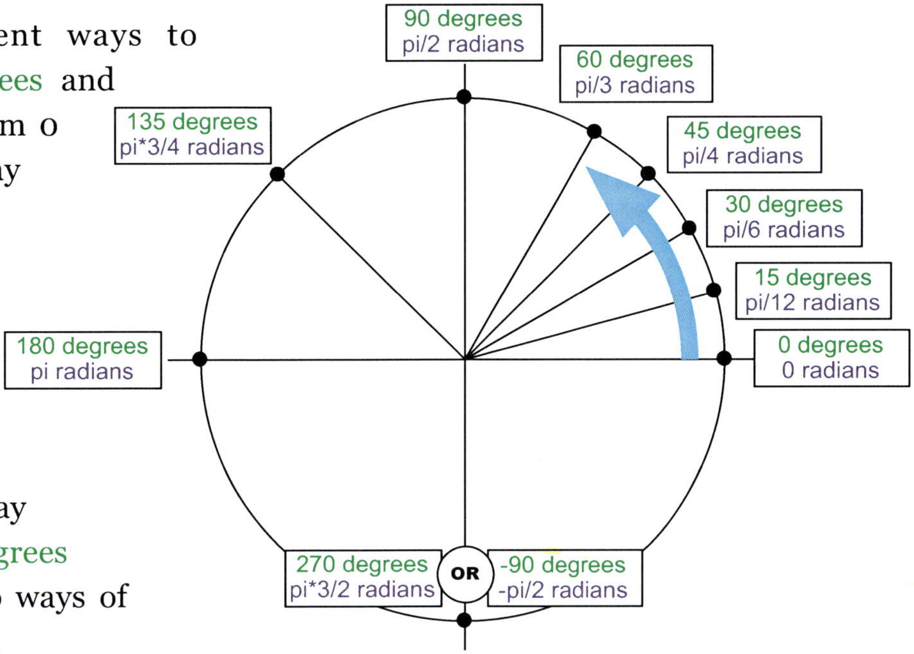

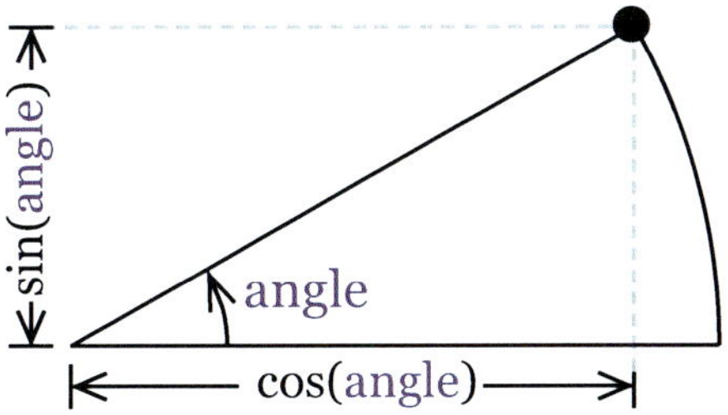

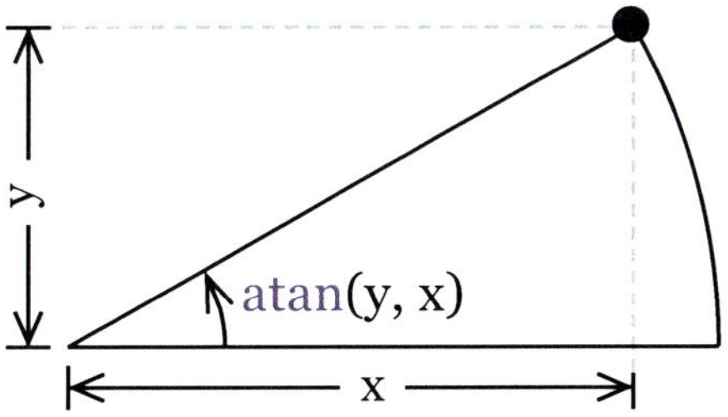

Most things in Mini Micro (like Sprite.rotation) use degrees. But some really handy functions called `sin` (sine), `cos` (cosine), and `atan` (arctangent) let you find the x and y coordinates of any point on a circle, or find the angle made by any x and y coordinates. These functions all use radians.

The table below shows how you can use these to get from things you know to things you may need. You can see this math in use in lines 10-15 of the the *Light Bikes* game (p. 44).

If You Have...	But You Want...	Then Do:
Angle (`a`) in degrees, Radius (`r`), and Center (`cx`, `cy`)	Pixel position (`x`, `y`)	`x = cx + r * cos(a * pi/180)` `y = cy + r * sin(a * pi/180)`
Pixel position (`x`, `y`) and Center (`cx`, `cy`)	Angle (`a`) in degrees or Radius (`r`)	`a = atan(y-cy, x-cy) * 180/pi` `r = sqrt((x-cx)^2 + (y-cy)^2)`

Glossary of Programming Terms

bug: a mistake in a computer program.

code: instructions in a computer program. Sometimes also called "source code."

code editor: a text editor specialized for editing computer programs (code).

command: a word with special meaning to a computer; using it in a program, or at a command prompt, causes the computer to do something.

comment: a part of a program that is ignored by the computer. Comments are added to explain or clarify the code.

cursor: a blinking square or line that indicates that the computer is waiting for you to type something. Or, in the case of "mouse cursor," an arrow that indicates the position of the mouse on screen.

debugging: the process of figuring out what's wrong with a program, and fixing it.

degrees: a way of measuring rotations or angles. One full rotation is 360 degrees.

dot syntax: a way of referring to values or functions in a map, by using the map, followed by a period, followed by the name of the value or function. For example, `gfx.line` refers to the `line` function on the `gfx` object.

file: a document stored on the disk of a computer. Files can represent anything: images, sounds, programs, etc.

frames (of animation): a series of images intended to be presented in sequence, like an animated cartoon.

function: a sub-program. A function wraps up a whole section of code into something that can be run with a single command, wherever that is needed.

game jam: a type of contest where participants write a game in a short amount of time (often 3-7 days), based on a theme that is not announced until the contest begins.

histogram: a kind of bar chart where each bar represents a count of values within a certain range.

indexing: looking up a particular item in a list, string, or map by putting the index of the thing you want in square brackets. For example, `msg[3]` gets element 3 of `msg`.

infinite loop: a loop that repeats forever, until either the user stops it by pressing Control-C, or the code breaks out of the loop with `break`.

list: an ordered sequence of values. In MiniScript, these are represented with square brackets.

loop: a section of code that repeats multiple times.

map: a collection of key-value pairs. In MiniScript, maps are enclosed in curly braces. Maps also support *dot syntax*.

Mini Micro: a neo-retro virtual computer that lets you create fun games and demos in a uniform way, no matter what sort of real computer you have.

MiniScript: a programming language known for being especially simple, clear, and easy to learn, while still powerful enough to write real programs.

object: a kind of map that contains both values and functions.

operator: a symbol that goes into code to represent some operation on one or two values. For example, `+` is the "plus operator" and represents addition of numbers, or joining of strings or lists.

parameter: a value passed into a function. For example, in `val("42")`, `"42"` is the parameter.

pixel: short for picture element; a tiny square of color on a computer screen or image.

program: a series of instructions for a computer, written in some programming language.

programming language: a special language for writing instructions for a computer. MiniScript is one example.

prompt: a string or character that tells the user what the computer is waiting for them to type. In Mini Micro, the command prompt is `]`, which lets you know it is waiting for a command.

radians: another way of measuring rotations and angles. One full rotation is 2*pi radians.

random: something that gives a different result every time, like rolling a die, or using the `rnd` function.

sprite: a small image on a computer screen that can be efficiently moved, scaled, rotated, or given a color tint.

sprite sheet: an image designed to be carved up into multiple *frames* for use by a sprite (or several sprites).

string: an ordered set of letters, numbers, and punctuation such as `"Hello world!"` Messages in a computer program are always represented as strings. The term comes from a bunch of characters strung together in order, like beads on a string.

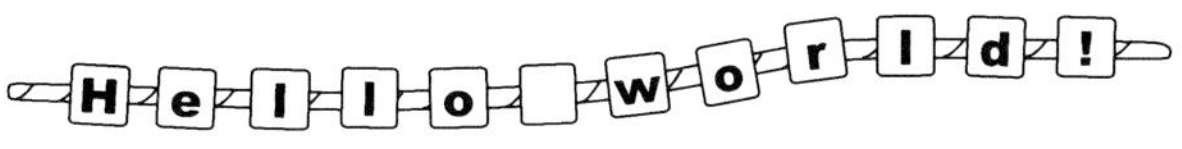

variable: a named storage space for a value in the computer's memory.

Going Further

This book has given you a taste of computer programming. If you want to learn more, there are many ways to do that!

One of the best ways is another book, *Learn to Code in 30 Days*. It will take you from step one to advanced programming skills in 30 short chapters.

If you go to https://miniscript.org, you will find links to other documentation, including the Mini Micro home page. Look for the *Mini Micro Cheat Sheet*, a 4-page quick reference that lists all the functions you can use in Mini Micro.

You can also join the MiniScript community online, which meets via Discord and web forums. The community is very supportive and always happy to help new coders. Find them via the website above.

Above all, keep experimenting and trying things. Computer programming is a deeply satisfying hobby, combining both logical and creative thinking. It can also lead to a variety of rewarding careers. But for now, just play around and have fun!

Index

Made in United States
North Haven, CT
07 November 2026

21928349R00033